DISCOVER PENGEOS YACHT THE FUTURE OF SEA LUXURY

TEDDY LONGMAN

CONTENTS

CHAPTER 1: INTRODUCTION

The Pangeos watercraft represents the epitome of maritime luxury and innovation, redefining the concept of a floating city. Designed to accommodate up to 60,000 guests, it offers a vast array of amenities and facilities to ensure an unparalleled experience at sea.

From world-class hotels and shopping centers to sprawling parks and recreational areas, Pangeos seamlessly blends the comforts of modern living with the tranquility of oceanic surroundings. Its state-of-the-art infrastructure includes ports for both ships and aircraft, facilitating seamless transportation to and from the vessel.

Whether guests seek relaxation, adventure, or cultural enrichment, Pangeos caters to every desire. Indulge in gourmet dining, unwind at luxurious spas, or partake in thrilling water sports—all against the backdrop of breathtaking sea views.

As an itinerant marvel, Pangeos traverses the oceans, offering travelers the opportunity to explore diverse destinations while enjoying unparalleled levels of comfort and sophistication. With its grandeur and versatility, Pangeos sets a new standard for luxury travel on the high seas.

CHAPTER 2: PROJECT PENGEOS

The ambitious project known as "Pangeos" aims to redefine maritime engineering and luxury with its proposed design a massive turtle-shaped vessel set to be an astonishing 550 meters long and 610 meters wide at its widest point. If realized, Pangeos would surpass all existing floating structures, earning the title of the largest terayacht ever constructed.

Proposed to be built in Saudi Arabia, the location offers an ideal environment for such an endeavor, with access to cutting-edge technology, skilled labor, and ample resources. The kingdom's strategic position on the Arabian Gulf provides a gateway to global shipping routes, facilitating the transportation of materials and components necessary for the construction of this monumental tera yacht.

Pangeos is envisioned as more than just a vessel; it is a testament to human ingenuity and ambition. Its turtle-inspired design not only captures the imagination but also serves a practical purpose, offering stability and efficiency in navigating the open seas.

Upon completion, Pangeos will boast a plethora of luxurious amenities and facilities, catering to the needs and desires of its privileged guests. From opulent staterooms and gourmet dining establishments to state-of-the-art entertainment venues and recreational areas, every aspect of Pangeos is designed to provide an unparalleled experience at sea.

Furthermore, the vessel's sheer size and grandeur will undoubtedly capture the attention of the world, solidifying Saudi Arabia's reputation as a hub for innovation and excellence in maritime engineering.

As plans for Pangeos move forward, it represents not only a remarkable feat of engineering but also a symbol of ambition and vision, destined to leave an indelible mark on the maritime landscape for generations to come.

The steel structure of the hull is composed of 30,000 individual compartments that serve the purpose of maintaining the buoyancy of Pangeos. The tera-turtle has the ability to navigate across the globe without interruption, as it does not produce any emissions due to its access to an essentially infinite source of sustainable energy. The exterior of the vessel is adorned with solar panels that capture and convert clean energy from the sun to power both the hotel facilities and the propulsion system. With the assistance of nine electric motors, each with a capacity of 16,800 horsepower, Pangeos is capable of reaching a top speed of 5 knots. Furthermore, the "wings" possess the capability to harness additional energy from both the waves and the wind.

The Pengeos Tera Yacht is a visionary project proposed to be built in Saudi Arabia, representing the pinnacle of maritime luxury and engineering. With an estimated cost of $8 billion, this floating city sets out to redefine the boundaries of extravagance and innovation on the open seas.

Conceptually, the Pengeos Terayacht embodies the fusion of advanced technology, exquisite design, and unparalleled luxury. Designed to resemble a majestic turtle, the vessel's awe-inspiring dimensions are staggering, projected to be 550 meters long and 610 meters wide at its widest point. If realized, it would undoubtedly claim the title of the largest floating structure ever constructed, setting a new standard for maritime excellence.

Situated in Saudi Arabia, the construction of the Pengeos Tera Yacht benefits from the kingdom's strategic location, robust infrastructure, and access to cutting-edge technology.

This ambitious undertaking represents a testament to Saudi Arabia's commitment to innovation and its aspirations to become a global leader in maritime engineering.

Beyond its impressive size, the Pengeos Terayacht promises to offer an unparalleled array of amenities and facilities, designed to cater to the most discerning of guests.

From luxurious hotels and shopping centers to sprawling parks and entertainment venues, every aspect of the floating city is meticulously planned to provide an unforgettable experience at sea.

Moreover, the Pengeos Terayacht is envisioned as more than just a symbol of opulence; it is a testament to sustainability and environmental stewardship. Incorporating state of the art eco-friendly technologies and practices, the vessel aims to minimize its carbon footprint and set new standards for responsible luxury travel.

CHAPTER 3: CONCLUSION

In conclusion, the Pengeos Terayacht stands as a testament to human ingenuity, ambition, and the relentless pursuit of excellence. With its visionary design, awe-inspiring dimensions, and unprecedented level of luxury, it represents a bold leap forward in maritime engineering and hospitality.

As plans for its construction move forward in Saudi Arabia, the Pengeos Terayacht embodies the kingdom's commitment to innovation and its aspirations to become a global leader in maritime excellence. The project not only showcases Saudi Arabia's capabilities in engineering and technology but also highlights its dedication to pushing the boundaries of what is possible on the open seas.

Beyond its grandeur and opulence, the Pengeos Terayacht symbolizes a new era of sustainable luxury travel. With its incorporation of eco-friendly technologies and practices, it sets a new standard for responsible tourism and environmental stewardship in the maritime industry.

As the world eagerly anticipates the realization of this monumental project, the Pengeos Terayacht serves as a beacon of inspiration a testament to what can be achieved through vision, innovation, and determination. It is poised to captivate the imagination of the world and leave an indelible mark on the maritime landscape for generations to come.

QUESTIONS

1. What makes the Pengeos Tera Yacht stand out as a visionary project in maritime engineering?

2. How does the proposed size of the Pengeos Tera Yacht compare to other floating structures?

3. What strategic advantages does Saudi Arabia offer for the construction of the Pengeos Tera Yacht?

4. How does the Pengeos Terayacht aim to redefine luxury travel on the open seas?

5. What are some of the key amenities and facilities that guests can expect aboard the Pengeos Tera Yacht?

6. In what ways does the Pengeos Terayacht demonstrate a commitment to sustainability and

environmental stewardship?

7. What role does innovation play in the development of the Pengeos Tera Yacht?

8. How does the Pengeos Terayacht reflect Saudi Arabia's aspirations to become a global leader in maritime excellence?

9. What impact would the construction of the Pengeos Tera Yacht have on the maritime

industry and luxury travel?

10. What factors contribute to the Pengeos Tera Yachts potential to captivate the imagination of the world?